Tyrannosaurus rex

The name Tyrannosaurus rex (say **ti-ran-o-sore-us**) or T rex means 'king of lizards'. The first part of its name comes from the Greek words for **'tyrant lizard'** and **'rex'** is from the Latin word for 'king'.

A T rex skeleton had about **200** bones, about the same number as humans, but they were MUCH BIGGER. Some bones had **holes** in them, to make the body lighter and easier to move around.

Meat-eating dinosaurs like T rex are called **theropods.**

Scientists believe T rex could eat up to 500lbs/230kg of flesh in one **BIG** bite! That's the same weight as 100 chickens.

T rex was a fierce hunter that hunted, killed, and **ate** other dinosaurs, and probably preyed on weak and sick animals.

Its long, heavy tail was made up of **SMALL BONES** called vertebrae.

T rex weighed about the same as TWO elephants!

T rex had two small, short front **ARMS**, each with two clawed **FINGERS**. It was a long stretch from its arms to its mouth!

How **fast** was T rex? Some experts say it was slow, and **waddled** like a duck. Others say it could run as fast as a car travels, **25mph/40kph.**

One of the **largest dinosaur teeth** ever found was that of a T rex. It was **1ft/30cm long**. Use a ruler to see just how **big** that is!

T rex had two hind (back) legs, and **walked on its toes.** It had **three clawed toes** at the front of each foot, and **a small one** at the back.

Dinosaurs were **reptiles** (cold-blooded animals) but they **DID NOT LOOK** like the reptiles on Earth today, which include **crocodiles**, turtles, and snakes.

It lived in river valleys covered in forests of trees in what is now the **western part of North America**.

T rex lived during what scientists call the **Cretaceous** period, about 66 MILLION YEARS ago.

Triceratops

LARGE HERDS of Triceratops probably lived together, moving from place to place to find FOOD.

The name Triceratops (say **try-sera-tops**) comes from the Greek language.

tri (3) + keratops (horned face) = triceratops (3-horned face)

Triceratops lived in the forests and marshes of **western North America**.

The first FOSSILS (dinosaur bones preserved as rock) were found in 1887 in Colorado, USA, and Triceratops is the STATE DINOSAUR of **Wyoming**.

Triceratops was not a hunter and did not need to move fast to catch prey. It moved quite **s-l-o-w-l-y** at about 10mph/16kph.

The fossil skeleton of **Triceratops Cliff** is on show at Boston Museum of Science, USA. **Cliff** cost 1 million dollars!

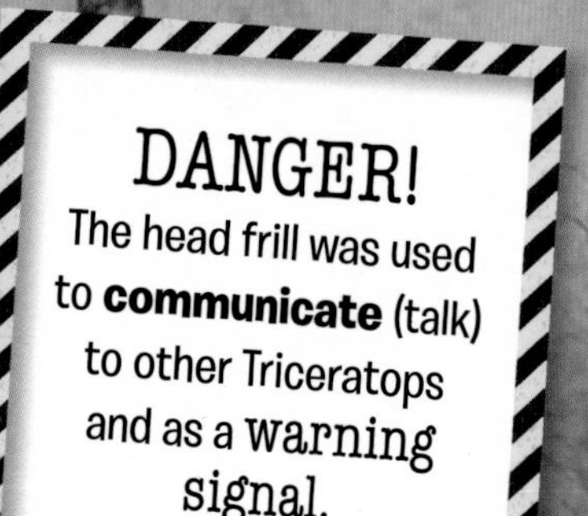

A large **HEAD FRILL** (a flap of tough skin) covered its neck like body armor. It protected against bites from predators like **T rex** and **Spinosaurus**, and in fights with other Triceratops. It was also protected by thick hide (skin).

Triceratops was a herbivore, a plant eater. It may have **knocked down trees** to get to the leaves.

Triceratops lived about **65 million** years ago.

Triceratops was about the size of an ELEPHANT, up to **30ft/9m** long, and **10ft/3m** tall. It weighed about **7 tons/6 tonnes**.

Triceratops had **3 horns** on its head. Two 'brow' horns above the eyes were about **3ft/1m long** and made of solid bone. A shorter horn on the snout (nose) was not bone, more like a **big fingernail.**

Triceratops had up to **800 teeth** – but not all at the same time! Arranged in groups, as some teeth wore out they were replaced by others. It had a mouth shaped like a **parrot's beak**, for pulling and crushing plants like ferns and palms.

Triceratops needed **big legs** to carry its **huge body**. Its fore (front) limbs (legs) were shorter than its rear (back) legs. It had **3 hooves** on its fore limbs and **4 hooves** on its rear limbs.

The **HUGE** head of Triceratops was one third the size (about 33%) of its whole body. SKULLS have been found that measure **8ft/2.5m.**

Pterosaurs

Flying Reptiles

Pterosaurs were as **tiny as a small bird** or as huge as a small plane!

Could dinosaurs fly? NO! Pterosaurs (say ter-o-saws) did fly, but they were FLYING REPTILES, not dinosaurs. Pterosaur means **'winged lizard'**.

Most Pterosaur **skulls** were full of tiny teeth, like sharp needles, but some had no teeth at all!

Pterosaur bones were not solid (like dinosaur bones). They were hollow (empty) and filled with air sacs (pockets). **WHY?** It made them light enough to lift their bodies into the air.

Most Pterosaurs had back-facing, cone-shaped head crests. What were they for? Displaying (showing off) to find a **mate?** Controlling the direction of their **flight?** Experts are not sure!

Pterosaurs did not have wings with **feathers**. Some had wings made of muscle and skin stretched between the VERY **l-o-n-g** 4th finger of their 'hand' and their back limb (leg).

Were Pterosaurs the first creatures to **fly?** NO - insects were the first.

Flying kept Pterosaurs **safe** from predators (hunters) but some were caught. How do we know? A Pterosaur fossil bone had a dinosaur tooth in it!

Pterosaurs flapped their wings but also used the wind to **GLIDE** through the air, using less energy.

Pterosaurs were carnivores (meat eaters). Most ate fish, flying low over rivers and seas to catch them in their **long, narrow jaws.** Some had throat 'pouches' (pockets) for scooping up fish, like today's pelicans. On land they ate insects and dead animals.

Pterosaurs were NOT designed for walking, and were **clumsy** on land.

Quetzalcoatlus (say kwet-zal-co-at-lus) may be the largest creature EVER TO FLY. It measured about 50ft/15m from one wing tip to the other, and its head was the size of a small **CAR!**

Pterosaurs lived in **North** and **South America**, and parts of **Europe** and **Asia** from over 200 MILLION years ago.

Pteranodon (say ter-an-o-don) had **3 fingers** on the front of each wing, and short back legs.

Dinosaur limbs (legs) were under their bodies. Pterosaur limbs stuck out from their sides, like lizards and crocodiles.

Pteranodon Café
MENU
fish
dead dinosaur bits
crabs
insects

Iguanodon

Iguanodon (say **ig-wah-no-don**) was a herbivore, a plant-eating dinosaur. It lived about 125 MILLION years ago on mainland Europe and in England, North America, and Africa.

A great discovery was made near a coal mine in Belgium in 1878 – the fossil bones of more than **30 Iguanodons!**

Iguanodon was large and **bulky**, not built for speed. But its size made smaller meat eaters less likely to attack it.

How big is big?

Iguanodon's front limb was about the size of a man and it weighed about as much as a van!

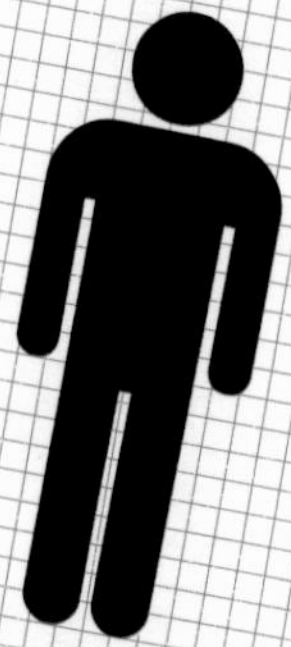

Iguanodon had a **sharp spike** for a thumb and used it to stab leaves and open seeds. It may have used it to stab hunters that wanted to eat it, too!

★ TOP FACTS

IGUANO-FACTS

How long?	about 30ft/10m
How heavy?	about 3 tons/3 tonnes
How fast?	about 12mph/20kph

Iguanodons laid eggs on land, sometimes buried in sand. Babies hatched out, like **birds**.

What we know about dinosaurs can **change**. We first thought Iguanodons walked on all **4 limbs**, then that they walked on **2 back limbs**. In fact, they may have walked on **2 OR 4 limbs!**

Some people think **Iguanodon** had **83 teeth!**

Adult humans have **32!**

thumb!

When **experts** first found Iguanodon **fossil bones** they built the skeleton with the **thumb spike** on the **NOSE!**

Iguanodon had a very long **5th finger**. It used it to gather plant food like **leaves, shoots,** and **twigs**.

Iguanodons had powerful back limbs (legs) and shorter front ones, with small hooves. Each foot had **3 toes.**

Iguanodons may have lived in **herds** (groups). Why do we think this? Because lots of **skeletons** have been found together in **one place.** Living in herds protected them from **hunters.**

Chewing tough plants all day was hard on Iguanodon **teeth!** As they wore out they were **replaced** one by one by new ones.

Diplodocus

DIPLODOCUS (say **dip-lod-ic-us**) was a dinosaur called a sauropod. It lived in the western part of North America about 150 MILLION years ago.

Dinosaurs lived **millions of years ago** but we have only known about them for about the last 200 years, since the 1820s.

Diplodocus was about 100ft/30m long. Its neck measured **20ft/6m** and its tail was EVEN **L-O-N-G-E-R**, with about **80 bones!** That's longer than a **bus!**

Diplodocus's brain may have been just **4ins/10cm** long! It MAY have had a 'second brain', a swelling near the base of its tail that helped control it.

Diplodocus swallowed stones that stayed in its stomach. They helped to grind the hard leaves it ate into a softer pulp.

Diplodocus lived on land, in groups called **herds**, but could wade through water in lakes and rivers to find FOOD.

Diplodocus had **5 toes** on each foot and **1 large claw** on the first toe of each back foot.

Diplodocus may have been the
l-o-n-g-e-s-t
dinosaur measured **head** to **tail.**
We know this because skeletons have been found and measured. Other dinosaurs may have been longer, but we cannot be sure as **no complete skeletons** have been found.

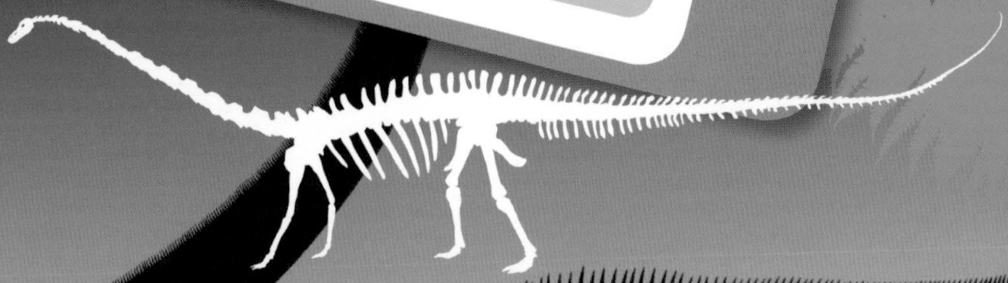

Thwack!
Diplodocus may have used its tail as a **weapon**.

Diplodocus
ate plants and trees like **mosses, ferns, cycads, gingkoes, and conifers**.

Diplodocus
was slim and light (about **11-13 tons/10-12 tonnes**) compared to other dinosaurs. **Brachiosaurus** weighed about TWICE as much!

Because it was a herbivore **(a plant eater)** Diplodocus's little 'peg' teeth were all at the front of its mouth, to rip leaves from plants. It had no back teeth because it did not chew meat. Its teeth may have lasted for only about a month!

Fossil **footprints** left in **mud** tell us that Diplodocus walked on **4 big feet**, a bit like an elephant. It moved quite s-l-o-w-l-y at about **5mph/8kph**, but could move faster if it was in danger.

Fossils of Diplodocus were first found in **Colorado, USA**, in 1877.
They have also been found in the states of **Montana** and **Utah**.

Plesiosaurs
Water Reptiles

Some **Plesiosaurs** were only about **10ft/3m** long.

Plesiosaurs (say ples-i-a-saws) were **water reptiles that lived** on Earth about 65 million years ago. They swam in both **fresh water** in lakes and rivers, and in **salt water** in seas and oceans.

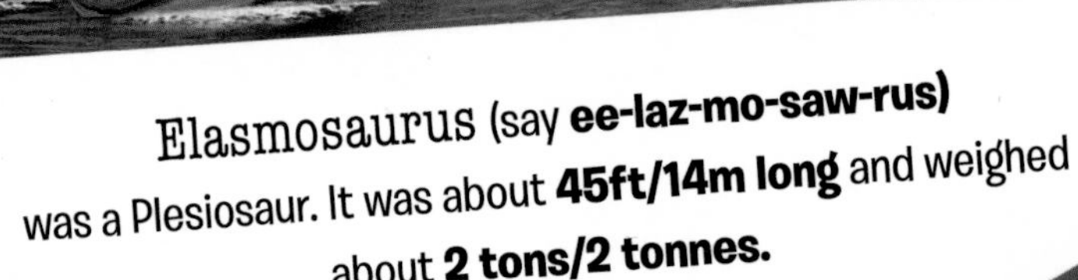

Elasmosaurus (say **ee-laz-mo-saw-rus)** was a Plesiosaur. It was about **45ft/14m long** and weighed about **2 tons/2 tonnes.**

The first **Elasmosaurus** fossils found were **VERY different** from anything seen before! When experts fitted the skeleton together they put its head on the end of its **tail!**

Each of Elasmosaurus's **4 paddle-shaped flippers** may have been about the size of a **man!**

Elasmosaurus's **l-o-n-g** neck was half of its whole length and had about **70 bones** in it!

Plesiosaurs swam under water to **hunt for food.** They had strong jaws, and ate bony fish, and animals like today's water snails, squid, and octopus.

Parts of a skeleton were the first Plesiosaur fossils (remains) found in England in 1719.

Question:
Did Plesiosaurs walk on land?
Answer:
NO!
Because they had flippers, not legs, they **DID NOT** walk on land.

The huge bodies of Plesiosaurs helped keep them safe from attack. So did their **sharp teeth!**

There are 2 kinds of Plesiosaur:

True Plesiosaur
long neck, small head

Pliosaur
shorter neck, large head

★ TOP FACTS

Some people believe these 'monsters' are Plesiosaurs still living in lakes and lochs!

name	place
Nessie	Loch Ness, Scotland
Champ	Lake Champlain, North America
Ogopogo	Lake Okanagan, Canada

Nessie?

Plesiosaurs had a broad body, a short tail, and 4 swimming flippers instead of limbs (legs).

Plesiosaurs MAY have laid eggs in sand at the water's edge, like today's **turtles**.

Stegosaurus

Stegosaurus (say **steg-uh-saw-rus**) was a large dinosaur that lived in the **Late Jurassic** period about 150 MILLION years ago.

Stegosaurus's TAIL SPIKES were up to 3ft/90cm long. Some had **2 sets of 2 spikes**, and some had more.

Stegosaurus had a HUGE body but a TINY brain. It may have been as small as a golf ball!

★ TOP FACTS

As BIG AS A BUS! Stegosaurus was:

- about **30ft/9m** long
- about **10ft/3m** tall
- about **5.5 tons/5 tonnes** in weight

Stegosaurus is the STATE DINOSAUR of Colorado, USA. The first fossils were found there in **1877**.

Stegosaurus lived in **herds** (groups) of both young and old animals. It was SAFER than living alone.

Plant eaters like Stegosaurus **did not have to run** to catch food. Large and **bulky**, it probably walked slowly, at around **4mph/7kph**.

ANTI-ATTACK!
Stegosaurus's size may have put off hunters like Allosaurus. So might its bony back plate armor, and slaps from its spiky tail!

Stegosaurus may have swallowed stones. In its stomach their rough edges helped **break up** the tough leaves it ate.

Othniel Marsh found the first Stegosaurus fossils. The bony back plates made him think of roof tiles, so he gave it a name from 2 Greek words:

stegos (roofed) + saurus (lizard) = stegosaurus

Stegosaurus had about **17** bony '**plates**' called SCUTES along its back. The largest measured **2ft/60cm** and may have been for display (showing off).

At the front of its mouth Stegosaurus had a **'beak'** - but no teeth! Its teeth were in its cheeks, and it used them for chewing.

Different kinds of Stegosaurus have been found in **Europe** (in **Portugal**), **China,** and **Africa**.

Stegosaurus was a herbivore that ate forest plants and trees like mosses, horsetails, and conifers. Did Stegosaurus eat grass? **NO** - grass DID NOT EXIST at that time!

Stegosaurus may have been able to **lift itself up** on to its **back legs** to reach leaves on tall trees.

Quiz

Try answering these questions to test your TOP FACT 100 info intake.
Answers below.

1. Did Iguanodon eat PLANTS or MEAT?

2. Which dinosaur's brain may have been as small as a **golf ball?**

3. Which dinosaur name means 'king of lizards'?

4. How many **horns** did Triceratops have on its head?
   - a 1
   - b 2
   - c 3

5. Pterosaurs could fly. **True** or **False?**